Table of Contents

What is RFID?

RFID tags are classed according to the radio frequency spectrum in which they communicate (low, high, or ultra-high) and how they communicate with the reader (active or passive).

Low frequency (LF), high frequency (HF), and ultra-high frequency (UHF) RFID tags are classified according to the frequency ranges they use to transmit data. In general, the lower the RFID system's frequency, the shorter the read range and the slower the data read rate.

Standard:	Description:
ISO 11784/11785	Animal identification. Uses 134.2 kHz.
ISO 14223	Radiofrequency identification of animals - Advanced transponders
ISO/IEC 14443	This standard is a popular HF (13.56 MHz) standard for HighFIDs which is being used as the basis of RFID-enabled passports under ICAO 9303. The Near Field Communication standard that lets mobile devices act as RFID readers/transponders is also based on ISO/IEC 14443.
ISO/IEC 15693	This is also a popular HF (13.56 MHz) standard for HighFIDs widely used for non-contact smart payment and credit cards.
ISO/IEC 18000	Information technology—Radio frequency identification for item management
ISO/IEC 18092	Information technology—Telecommunications and information exchange between systems—

	Near Field Communication— Interface and Protocol (NFCIP-1)
ISO 18185	This is the industry standard for electronic seals or "e-seals" for tracking cargo containers using the 433 MHz and 2.4 GHz frequencies.
ISO/IEC 21481	Information technology— Telecommunications and information exchange between systems—Near Field Communication Interface and Protocol -2 (NFCIP-2)
ASTM D7434	Standard Test Method for Determining the Performance of Passive Radio Frequency Identification (RFID) Transponders on Palletized or Unitized Loads
ASTM D7435	Standard Test Method for Determining the Performance of Passive Radio Frequency Identification (RFID)Transponders on Loaded Containers
ASTM D7580	Standard Test Method for Rotary Stretch Wrapper Method for Determining the Readability of Passive RFID Transponders on Homogenous Palletized or Unitized Loads
ISO 28560-2	Specifies encoding standards and data model to be used within libraries.

Active

Onboard active RFID tags have their own transmitter and power supply (often a battery). The majority of these are UHF solutions, with read ranges of up to 100 meters in some cases. Active tags, which are used to track major assets, are typically larger and more expensive than passive tags (like cargo containers, vehicles, and machines). Sensors on active RFID tags measure and transmit temperature, humidity, light, and shock/vibration data for the items to which they are connected.

Passive

The reader and reader antenna deliver a signal to the tag, which is used to power on the tag and reflect energy back to the reader in passive RFID solutions. Passive LF, HF, and UHF systems are available. The read range is lower than with active tags and is limited by the power of the reflected radio signal (commonly referred to as tag backscatter).

Battery Assisted Passive (BAP)

A hybrid tag is a third form of RFID tag. A power supply is included into a passive tag design in BAP systems, or semi-passive RFID systems.
The power supply ensures that all the reader's acquired energy is utilized to reflect the signal, resulting in increased read distance and data transfer rates. BAP tags, unlike active RFID transponders, lack their own transmitters.

Low Frequency (LF)

While low frequency technically includes frequencies between 30 to 300 KHz, only 125 KHz and 134.2 KHz are used in RFID communications. The read/write range of LF RFID tags is about 4 inches or 10 cm. Common standards for LF RFID include ISO 14223 and ISO/IEC 18000-2.

LF RFID Tag Properties	
Tag Type	Passive
Technology	Inductive Coupling
Frequency	125 KHz / 134 KHz
Read Rage	Short (few cm to Inches)
Read Speed	Slow
Works with Metal	Yes
Works with Liquids	Yes
Cost	Low

HID ProxCard (125 kHz)

HID prox cards are simply proximity cards branded by HID, a global pioneer in proximity cards and access management. An HID card, like other proximity and RFID cards, is essentially an ID card that allows proximity technology to be used in everyday operations.

EM4100x (125 kHz)

EM4100 compatible RFID transponders carry 64 bits of Read Only memory. This means that information can be read from the Tag, but no data can be changed, or new data written to the card once the card has been programmed with the initial data. When the Tag enters the electromagnetic field transmitted by the RFID reader it draws power from the field and will commence transmitting its data.

Indala (125 kHz)

The HID Indala 125 kHz Prox series of credentials use FlexSecur technology. This technology programs the cards and readers as a unique set for each customer so that one customer's reader can not read another customer's cards. In fact, Indala proximity readers can screen out unauthorized cards before sending data to the host system thus providing you with an extra layer of security and stopping would-be attackers at the first hurdle.

Hitag (125 kHz)

HITAG 1 based transponders are highly integrated and do not need any additional components beside the external coil.
Data between Key (RWD) and transponder is transmitted bidirectionally, in Half Duplex Mode. The HITAG 1 transponder IC offers also an encrypted data transmission.

The AntiCollision (AC) Mode, which is used mainly in long range operation, allows to handle several transponders that are at the same time in the communication field of the antenna, thus achieving highest operating reliability and permitting to handle several transponders quickly and simultaneously.

The HITAG 1 transponder IC provides two protocol modes, Standard and Advanced Mode. The Advanced Protocol Mode operates compared to the Standard Protocol Mode with an increased number of Startbits and a 8-bit Cyclic Redundancy Check (CRC) sent by the transponder IC at read operations. HITAG 1 transponder IC offer a memory of 2 kbit.

High Frequency (HF)

High frequency includes frequencies between 3 to 30 MHz, only 13.56 is used for RFID communication. The read range of HF tags is about 3 feet or 1 meter. Standards include ISO 15693, ECMA-340, ISO/IEC 18092 (for NFC), ISO/IEC 14443A and ISO/IEC 14443 (for MIFARE and other smart card solutions).

HF RFID Tag Properties	
Tag Type	Passive
Frequency	13.56 MHz (1.75 MHz to 13.56 MHz)
Read Rage	10 cm to 1 meter
Read Speed	Slow/Medium
Read Multiple Tags Simultaneously	Yes
Works with Metal	Yes
Works with Liquids	Yes
Cost	Low

HID iClass (13.56 MHz)

A proprietary technology made by HID, iCLASS® cards combine magnetic stripe functionality with a contactless smart card. Based on RFID, these smart cards use a high frequency radio technology, versus the low frequencies used in proximity cards

MIFARE Classic (13.56 MHz)

The MIFARE Classic IC is just a memory storage device, where the memory is divided into segments and blocks with simple security mechanisms for access control. They are ASIC-based and have limited computational power. Due to their reliability and low cost, those cards are widely used for electronic wallets, access control, corporate ID cards, transportation, or stadium ticketing.

Ultra-High Frequency (UHF)

The Ultra High Frequency (UHF) band spans 300 MHz to 1 GHz in frequency. The most common frequencies used by UHF are 433 MHz and 860 to 960 MHz.

UHF RFID Tag Properties	
Tag Type	Passive / Battery Assisted / Active
Frequency	433 MHz and 860 to 960 MHz
Read Range	Up to 50 feet
Read Speed	Very Fast
Read Multiple Tags Simultaneously	Yes
Works with Metal	No
Works with Water	No
Cost	Very Low

Cloning

Proxmark3

To use the ProxMark3 you will need to install software that allows for it to communicate with the operating system, below are the instructions for Windows, Linux and MacOS.

Install Software

Windows

```
1.   Download the latest release of ProxSpace from
     https://github.com/Gator96100/ProxSpace/releases and
     download the latest release .zip file.
2.   Extract to directory. The path must not have any spaces.
     (EX. C:\ProxSpace or C:\projects\proxspace)
3.   Run "runme64.bat"
4.   Run "git clone
     https://github.com/RfidResearchGroup/proxmark3.git"
5.   cd proxmark3
6.   make clean && make all
7.   ./pm3
```

Ubuntu Linux

```
sudo apt-get update

sudo apt-get install --no-install-recommends git ca-
certificates build-essential pkg-config \libreadline-dev gcc-
arm-none-eabi libnewlib-dev qtbase5-dev libbz2-dev
libbluetooth-dev

git clone https://github.com/RfidResearchGroup/proxmark3.git

sudo apt remove modemmanager

cd proxmark

make accessrights

sudo shutdown -r now
```

Fedora Linux

```
sudo dnf install git make gcc gcc-c++ arm-none-eabi-gcc-cs arm-
none-eabi-newlib readline-devel bzip2-devel qt5-qtbase-devel
bluez-libs-devel libatomic

git clone https://github.com/RfidResearchGroup/proxmark3.git

yum -y remove ModemManager

cd proxmark

make accessrights

sudo shutdown -r now
```

ArchLinux

```
sudo pacman -Sy git base-devel readline bzip2 arm-none-eabi-gcc
arm-none-eabi-newlib qt5-base bluez -needed

git clone https://github.com/RfidResearchGroup/proxmark3.git

sudo pacman -R modemmanager

cd proxmark

make accessrights

sudo shutdown -r now
```

openSUSE

```
sudo zypper install git patterns-devel-base-devel_basis gcc-c++
readline-devel libbz2-devel cross-arm-none-gcc9 cross-arm-none-
newlib-devel libqt5-qtbase-devel

git clone https://github.com/RfidResearchGroup/proxmark3.git
```

```
systemctl disable ModemManager.service

cd proxmark

make accessrights

sudo shutdown -r now
```

MacOS

The instructions below utilize homebrew, make sure to install homebrew for these instructions to work.

To install homebrew:

```
/bin/bash -c "$(curl -fsSL
https://raw.githubusercontent.com/Homebrew/install/HEAD/install
.sh)
```

```
brew install xquartz
brew tap RfidResearchGroup/proxmark3
brew install proxmark3

 To run client:

pm3
```

Tune HW

Command	Description
hw help	This help
hw detectreader	Detect external reader field
hw fpgaoff	Set FPGA off
hw lcd	Send command/data to LCD
hw lcdreset	Hardware reset LCD
hw readmem	Read memory at decimal address from flash
hw reset	Reset the Proxmark3
hw setlfdivisor	Drive LF antenna at 12Mhz
hw setmux	Set the ADC mux to a specific value
hw tune	Measure antenna tuning
hw version	Show version information about the connected Proxmark

LF

Identify Low Frequency (LF) Card Protocol:

```
[usb] pm3 --> lf search
```

Command	Description
[usb] pm3 --> lf help	This help
[usb] pm3 --> lf cmdread	Modulate LF reader field to send command before read (all periods in microseconds) (option 'h' for 134)
[usb] pm3 --> lf config	Set config for LF sampling, bit/sample, decimation, frequency
[usb] pm3 --> lf flexdemod	Demodulate samples for FlexPass
[usb] pm3 --> lf read	Read 125/134 kHz LF ID-only tag. Do 'lf read h' for help
[usb] pm3 --> lf search	Read and Search for valid known tag (in offline mode it you can load first then search) - 'u' to search for unknown tags
[usb] pm3 --> lf sim	Simulate LF tag from buffer with optional GAP (in microseconds)
[usb] pm3 --> lf simask	Simulate LF ASK tag from demodbuffer or input
[usb] pm3 --> lf simfsk	Simulate LF FSK tag from demodbuffer or input
[usb] pm3 --> lf simpsk	Simulate LF PSK tag from demodbuffer or input

[usb] pm3 --> lf simbidir	Simulate LF tag (with bidirectional data transmission between reader and tag)
[usb] pm3 --> lf snoop	Snoop LF (use lf config to set parameters) (needs a demod and/or plot after)
[usb] pm3 --> lf vchdemod	Demodulate samples for VeriChip (decimate first)

HID

Read and Convert	
pm3 --> lf hid read	Read HID Prox card
pm3 --> lf hid demod	Demodulate HID Prox card
pm3 --> lf hid wiegand [OEM] [FC] [CN]	Convert Facility code & Card number to Wiegand
Simulate HID	
weigand format	Simulate raw HID number [XXXXXXXXXX]
pm3 --> lf hid sim -w [XXXXXX] --fc [YY] --c [ZZZZ]	Simulate Wiegand with facility code and card numb
Clone HID	
pm3 --> lf hid clone - [XXXXXXXXXX]	Clone raw HID number [XXXXXXXXXX]
pm3 --> lf hid clone -w [XXXXXX] --fc [YY] --c [ZZZZ]	Clone Wiegand with facilit code and card number
Brute Force	
pm3 --> lf hid brute - [XXXXXX] -f [YYY]	Brute Force HID with facili code [X] and facility code [Y]
pm3 --> lf hid brute - [XXXXXX] -f [YYY] -c [ZZZ]	Brute Force HID with facili code [X] and facility code [Y] and [Z] code to start with
Options	
-v, --verbose	verbose logging

-w, --wiegand [format]	weigand format
-f, --fn [decimal]	facility code
-c, --cn [decimal]	card number to start with
-i [decimal]	issue level
-o, --oem [decimal]	OEM code
-d, --delay [decimal]	delay between attempts in m
--up	direction to increment card number.
--down	direction to increment card number.

Indala

Read & Demod Indala	
pm3 --> lf indala read	Read Indala Card
pm3 --> lf indala demod	Demodulate Indala card
pm3 --> lf indala altdemod	Alternative method to Demodulate samples for Indala 64 bit UID
Simulate Indala Card	
pm3 --> lf indala sim -r [XXXXXXXXXXXXXXXX]	Simulate raw bytes Indala card [XXXXXXXXXXXXXXXX]
Clone Indala Card to T5577 Card	
pm3 --> lf indala clone -r [XXXXXXXXXXXXXXXX]	Clone raw Indala card [XXXXXXXXXXXXXXXX]
Options	
-r, --raw <hex>	raw bytes

--heden [decimal]	Cardnumber for Heden 2L format
--fc [decimal]	Facility Code (26 bit H10301 format)
--cn [decimal]	Cardnumber (26 bit H10301 format)
--q5	specify writing to Q5/T5555 ta
--em	specify writing to EM4305/4469 tag

T55XX

Detect & Modulate T55XX	
pm3 --> lf t55xx detect	Detect T55XX card
pm3 --> lf t55xx config --[option]	Configuration demodulation for T55XX with [option]
Write to T55XX Card	
pm3 --> lf t55xx write -b 0 -d [XXXXXXXX]	
Wipe T55XX Card	
pm3 --> lf t55xx wipe	Wipe a T55XX tag and set defaults
Set T55XX Card Timings	
pm3 --> lf t55xx deviceconfig -zp	Set T55XX Card Timings
Modulation Options	
--FSK	set demodulation FSK
--FSK1	set demodulation FSK 1
--FSK1A	set demodulation FSK 1a (inv)
-FSK2	set demodulation FSK 2
--FSK2A	set demodulation FSK 2a (inv)

--ASK	set demodulation ASK
--PSK1	set demodulation PSK 1
-PSK2	set demodulation PSK 2
-PSK3	set demodulation PSK 3
--NRZ	set demodulation NRZ
--BI	set demodulation Biphase
--BIA	set demodulation Diphase (inverted biphase)
Note*	
EM is ASK HID Prox is FSK Indala is PSK	
T55XX Write Options	
-b, --blk <0-7>	block number to write
-d, --data <hex>	data to write (4 hex bytes)
-p, --pwd <hex>	password (4 hex bytes)
Timing Options	
-p, --persist	persist to flash memory (RDV4)
-z	Set default t55x7 timings (use `-p` to save if required)

Command	Description
lf t55xx help	This help
lf t55xx bruteforce	Simple brute force attack to find password
lf t55xx config	Set/Get T55XX configuration (modulation, inverted, offset, rate)
lf t55xx detect	Try detecting the tag modulation from reading the configuration block.

lf t55xx p1detect	Try detecting if this is a t55xx tag by reading page 1
lf t55xx read	Read T55xx block data (page 0)
lf t55xx resetread	Send Reset chip Cmd then lf read the stream to attempt to identify the start of it (needs a demod and/or plot after)
lf t55xx write	Write T55xx block data ([1] for page 1)
lf t55xx trace	Show T55xx traceability data (page 1/ blk 0-1)
lf t55xx info	Show T55xx configuration data (page 0/ blk 0)
lf t55xx dump	Dump T55xx card block 0-7. [optional password]
lf t55xx special	Show block changes with 64 different offsets
lf t55xx wakeup	Send AOR wakeup command
lf t55xx wipe	Wipe a T55xx tag and set defaults (will destroy any data on tag)

Hitag

Hitag Information	
pm3 --> lf hitag info	Read Hitag information
Act as a Hitag Reader	
pm3 --> lf hitag --01	HitagS, read all pages, challenge mode
pm3 --> lf hitag --02	HitagS, read all pages, crypto mode. Set key=0 for no auth
pm3 --> lf hitag --21 -k [XXXXXXXX]	Hitag2, read all pages, password mode. def 4D494B52 (MIKR)
pm3 --> lf hitag --22	Hitag2, read all pages, challenge mode

pm3 --> lf hitag reader --23 -k [XXXXXXXXXXXX]	Hitag2, read all pages, crypto mode. Key ISK high + ISK low. def 4F4E4D494B52 (ONMIKR)
pm3 --> lf hitag --25	Hitag2, test recorded authentications (replay?)
pm3 --> lf hitag --26	Hitag2, read UID
Sniff Hitag traffic	
pm3 --> lf hitag sniff	
pm3 --> lf hitag list	
Simulate Hitag	
pm3 --> lf hitag sim -2	
Write Hitag	
pm3 --> lf hitag writer --03	HitagS, write page, challenge mode
pm3 --> lf hitag writer --04	HitagS, write page, crypto mode. Set key=0 for no auth
pm3 --> lf hitag writer --24	Hitag2, write page, crypto mode
pm3 --> lf hitag writer --27	Hitag2, write page, password mode
Options	
-p, --page [dec]	page address to write to
-d, --data [hex]	data, 4 hex bytes
-k, --key [hex]	key, 4 or 6 hex bytes
--nrar [hex]	nonce / answer writer, 8 hex bytes

Command	Description
lf hitag help	This help
lf hitag list	List Hitag trace history

lf hitag reader	Act like a Hitag Reader
lf hitag sim	Simulate Hitag transponder
lf hitag snoop	Eavesdrop Hitag communication
lf hitag writer	Act like a Hitag Writer
lf hitag simS	<hitagS.hts> Simulate HitagS transponder
lf hitag checkChallenges	<challenges.cc> test all challenges

EM4X

Command	Description
lf em help	This help
lf em 410xread	Extract ID from EM410x tag in GraphBuffer
lf em 410xdemod	Extract ID from EM410x tag (option 0 for continuous loop, 1 for only 1 tag)
lf em 410xsim	Simulate EM410x tag
lf em 410xwatch	Watches for EM410x 125/134 kHz tags
lf em 410xspoof	Watches for EM410x 125/134 kHz tags, and replays them.
lf em 410xwrite	Write EM410x UID to T5555(Q5) or T55x7 tag, optionally setting clock rate
lf em 4x05dump	Read all EM4x05/EM4x69 words
lf em 4x05info	Read and output EM4x05/EM4x69 chip info
lf em 4x05readword	Read EM4x05/EM4x69 word data

lf em 4x05writeword	Write EM4x05/EM4x69 word data
lf em 4x50read	Extract data from EM4x50 tag

fdx

Command	Description
lf fdx help	This help
lf fdx demod	Attempt to extract FDX-B ISO11784/85 data from the GraphBuffer
lf fdx read	Attempt to read and extract FDX-B ISO11784/85 data
lf fdx sim	Animal ID tag simulator
lf fdx clone	Clone animal ID tag to T55x7 (or to q5/T5555)

G Prox II

Command	Description
lf gproxii help	This help
lf gproxii demod	Demodulate a G Prox II tag from the GraphBuffer
lf gproxii read	Attempt to read and Extract tag data from the antenna

IO Prox

Command	Description

lf io help	This help
lf io demod	Demodulate IO Prox tag from the GraphBuffer
lf io read	Realtime IO FSK demodulator
lf io clone	Clone ioProx Tag

JABLOTRON

Command	Description
lf jablotron help	This help
lf jablotron demod	Attempt to read and extract tag data from the GraphBuffer
lf jablotron read	Attempt to read and extract tag data from the antenna
lf jablotron clone	clone jablotron tag
lf jablotron sim	simulate jablotron tag

NexWatch

Command	Description
lf nexwatch help	This help

lf nexwatch demod	Demodulate a NexWatch tag from the GraphBuffer
lf nexwatch read	Attempt to read and extract tag data from the antenna

Noralsy

Command	Description
lf noralsy help	This help
lf noralsy demod	Attempt to read and extract tag data from the GraphBuffer
lf noralsy read	Attempt to read and extract tag data from the antenna
lf noralsy clone	clone Noralsy tag
lf noralsy sim	simulate Noralsy tag

Paradox

Command	Description
lf paradox help	This help
lf paradox demod	Attempt to read and extract tag data from the GraphBuffer

lf paradox read	Attempt to read and extract tag data from the antenna
lf paradox clone	Clone Paradox to T55x7

PCF7931

Command	Description
lf pcf7931 help	This help
lf pcf7931 read	Read content of a PCF7931 transponder
lf pcf7931 write	Write data on a PCF7931 transponder.
lf pcf7931 config	Configure the password, the tags initialization delay and time offsets

Presco

Command	Description
lf presco help	This help
lf presco read	Attempt to read and extract tag data from the antenna
lf presco clone	d or h clone presco tag
lf presco sim	d or h simulate presco tag

Pyramid

Command	Description
lf pyramid help	This help
lf pyramid demod	Attempt to read and extract tag data from the GraphBuffer
lf pyramid read	Attempt to read and extract tag data from the antenna
lf pyramid clone	clone pyramid tag
lf pyramid sim	simulate pyramid tag

Securakey

Command	Description
lf securakey help	This help
lf securakey demod	Attempt to read and extract tag data from the GraphBuffer
lf securakey read	Attempt to read and extract tag data from the antenna

TI

Command	Description
lf ti help	This help

lf ti demod	Demodulate raw bits for TI-type LF tag
lf ti read	Read and decode a TI 134 kHz tag
lf ti write	Write new data to a r/w TI 134 kHz tag

Viking

Command	Description
lf viking help	This help
lf viking demod	Attempt to read and extract tag data from the GraphBuffer
lf viking read	Attempt to read and extract tag data from the antenna
lf viking clone	clone viking tag
lf viking sim	Simulate viking tag

Visa2000

Command	Description
lf visa2000 help	This help
lf visa2000 demod	Attempt to read and extract tag data from the GraphBuffer

lf visa2000 read	Attempt to read and extract tag data from the antenna
lf visa2000 clone	clone Visa2000 tag
lf visa2000 sim	simulate Visa2000 tag

Cotag

Command	Description
lf cotag help	This help
lf cotag demod	Tries to decode a COTAG signal
lf cotag read	Attempt to read and extract tag data

AWID

Command	Description
lf awid help	This help
lf awid demod	Demodulate an AWID FSK tag from the GraphBuffer
lf awid read	Realtime AWID FSK read from the antenna
lf awid sim	AWID tag simulator
lf awid clone	Clone AWID to T55x7 (tag must be in range of antenna)

HF

Identify High Frequency (HF) Card Protocol:

```
[usb] pm3 --> hf search
```

Command	Description
hf help	This help
hf tune	Continuously measure HF antenna tuning
hf list	List protocol data in trace buffer
hf search	Search for known HF tags (identifies 14443a, 14443b, 15693, iClass)

ICLASS

Operations	
Command	**Description**
help	This help
dump	Dump Picopass / iCLASS tag to file
info	Tag information
list	List iclass history
rdbl	Read Picopass / iCLASS block
reader	Act like an Picopass / iCLASS reader
restore	Restore a dump file onto a Picopass / iCLASS tag
sniff	Eavesdrop Picopass / iCLASS communication
wrbl	Write Picopass / iCLASS block
Recovery	
Command	**Description**
chk	Check keys
loclass	Use loclass to perform bruteforce reader attack
lookup	Uses authentication trace to check for key in dictionary file
Simulation	
Command	**Description**
sim	Simulate iCLASS tag
eload	Load Picopass / iCLASS dump file into emulator memory
esave	Save emulator memory to file
eview	View emulator memory
Utils	
Command	**Description**
configcard	Reader configuration card

calcnewkey	Calc diversified keys (blocks 3 & 4) to write
encode	Encode binary wiegand to block 7
encrypt	Encrypt given block data
decrypt	Decrypt given block data or tag dump file
managekeys	Manage keys to use with iclass commands
permutekey	Permute function from 'heart of darkness'paper
view	Display content from tag dump file

MIFARE

Check for default keys	
pm3 --> hf mf chk --mini -f mfc_default_keys	MIFARE Classic Mini / S20 default keys from file mfc_default_keys
pm3 --> hf mf chk --1k -f mfc_default_keys	MIFARE Classic 1k / S50 default keys from file mfc_default_keys
pm3 --> hf mf chk --2k -f mfc_default_keys	MIFARE Classic/Plus 2k default keys from file mfc_default_keys
pm3 --> hf mf chk --4k -f mfc_default_keys	MIFARE Classic 4k / S70 default keys from file mfc_default_keys
Check for default keys from local memory	
hf mf fchk --mini --mem	
hf mf fchk --1k --mem	
hf mf fchk --2k --mem	
hf mf fchk --4k --mem	
Dump MIFARE Classic card contents	
pm3 --> hf mf dump --mini -k key.bin -f dump.bin	MIFARE Classic Mini / S20 contents with key.bin and file dump.bin
pm3 --> hf mf dump --1k -k key.bin -f dump.bin	MIFARE Classic 1k / S50 contents with key.bin and file dump.bin
pm3 --> hf mf dump --2k -k key.bin -f dump.bin	MIFARE Classic/Plus 2k contents with key.bin and file dump.bin
pm3 --> hf mf dump --4k -k key.bin -f dump.bin	MIFARE Classic 4k / S70 contents with key.bin and file dump.bin
Run autopwn, to extract all keys and backup a MIFARE Classic tag	

pm3 --> hf mf autopwn	run autopwn, extract keys and backup MIFARE classic
autopwn options	
-k, --key <hex>	Known key, 12 hex bytes
-s, --sector <dec>	Input sector number
-a	Input key A [default]
-b	Input key B
-f, --file <fn>	filename of dictionary
-s, --slow	Slower acquisition (required by some non standard cards)
-l, --legacy	legacy mode (use the slow `hf mf chk`)
-v, --verbose	verbose output (statistics)
--mini	MIFARE Classic Mini / S20
--1k	MIFARE Classic 1k / S50 [default]
--2k	MIFARE Classic/Plus 2k
--4k	MIFARE Classic 4k / S70
Simulate MIFARE	
pm3 --> hf mf sim -u [XXXXXXXX]	Simulate MIFARE with UID [XXXXXXXX]
Read MIFARE Ultralight EV1	
pm3 --> hf mfu info	Get MIFARE Ultralight information
Simulate MIFARE Sequence	
pm3 --> hf mf fchk --[type] -f mfc_default_keys.dic	MIFIRE default keys from file mfc_default_keys.dic
pm3 --> hf mf dump	MIFIRE dump
pm3 --> hf mf eload -f [uid-dump.bin]	Load MIFIRE from dump
pm3 --> hf mf sim -u [XXXXXXXX]	Simulate MIFARE with UID [XXXXXXXX]

MIFARE Specific Commands

Command	Description
hf mf dbg	Set default debug mode
hf mf rdbl	Read MIFARE classic block
hf mf rdsc	Read MIFARE classic sector
hf mf dump	Dump MIFARE classic tag to binary file
hf mf restore	Restore MIFARE classic binary file to BLANK tag
hf mf wrbl	Write MIFARE classic block
hf mf chk	Test block keys
hf mf mifare	Read parity error messages.
hf mf nested	Test nested authentication
hf mf sniff	Sniff card-reader communication
hf mf sim	Simulate MIFARE card
hf mf eclr	Clear simulator memory block
hf mf eget	Get simulator memory block
hf mf eset	Set simulator memory block
hf mf eload	Load from file emul dump
hf mf esave	Save to file emul dump

hf mf ecfill	Fill simulator memory with help of keys from simulator
hf mf ekeyprn	Print keys from simulator memory
hf mf csetuid	Set UID for magic Chinese card
hf mf csetblk	Write block - Magic Chinese card
hf mf cgetblk	Read block - Magic Chinese card
hf mf cgetsc	Read sector - Magic Chinese card
hf mf cload	Load dump into magic Chinese card
hf mf csave	Save dump from magic Chinese card into file or emulator

MIFIRE Ultralite Commands

Command	Description
hf mfu help	This help
hf mfu dbg	Set default debug mode
hf mfu info	Tag information
hf mfu dump	Dump Ultralight / Ultralight-C / NTAG tag to binary file
hf mfu rdbl	Read block
hf mfu wrbl	Write block
hf mfu cauth	Authentication - Ultralight C

hf mfu setpwd	Set 3des password - Ultralight-C
hf mfu setuid	Set UID - MAGIC tags only
hf mfu gen	Generate 3des mifare diversified keys

HF 14a

Command	Description
help	This help
list	List ISO 14443-a history
info	Tag information
reader	Act like an ISO14443-a reader
ndefread	Read an NDEF file from ISO 14443-A Type 4
cuids	Collect n>0 ISO14443-a UIDs in one go
sim	Simulate ISO 14443-a tag
sniff	sniff ISO 14443-a traffic
apdu	Send ISO 14443-4 APDU to tag
chaining	Control ISO 14443-4 input chaining
raw	Send raw hex data to tag
antifuzz	Fuzzing the anticollision phase. Warning
config	Configure 14a settings (use with caution)
apdufind	Enumerate APDUs - CLA/INS/P1P2

HF 14b

Command	Description
help	This help
apdu	Send ISO 14443-4 APDU to tag
dump	Read all memory pages of an ISO-14443-B tag, save to file
info	Tag information
list	List ISO-14443-B history
ndefread	Read NDEF file on tag
raw	Send raw hex data to tag
reader	Act as a ISO-14443-B reader to identify
sim	Fake ISO ISO-14443-B tag
sniff	Eavesdrop ISO-14443-B
rdbl	Read SRI512/SRIX4x block
sriwrite	Write data to a SRI512 or SRIX4K tag

HF 15

General	
help	This help
list	List ISO-15693 history
demod	Demodulate ISO-15693 from tag
dump	Read all memory pages of an ISO-15693 tag, save to file
info	Tag information

sniff	Sniff ISO-15693 traffic
raw	Send raw hex data to tag
rdbl	Read a block
rdmulti	Reads multiple blocks
reader	Act like an ISO-15693 reader
restore	Restore from file to all memory pages of an
samples	Acquire samples as reader (enables carrier,
sim	Fake an ISO-15693 tag
slixdisable	Disable privacy mode on SLIX ISO-15693 tag
wrbl	Write a block
afi	
findafi	Brute force AFI of an ISO-15693 tag
writeafi	Writes the AFI on an ISO-15693 tag
writedsfid	Writes the DSFID on an ISO-15693 tag
magic	
csetuid	Set UID for magic card

EPA

Command	Description
help	This help
cnonces	Acquire encrypted PACE nonces of specific size
preplay	Perform PACE protocol by replaying given APDUs

Legic

Command	Description
help	This help
list	List LEGIC history
reader	LEGIC Prime Reader UID and tag info
info	Display deobfuscated and decoded LEGIC Prime tag data
dump	Dump LEGIC Prime tag to binary file
restore	Restore a dump file onto a LEGIC Prime
rdbl	Read bytes from a LEGIC Prime tag
sim	Start tag simulator
wrbl	Write data to a LEGIC Prime tag
crc	Calculate Legic CRC over given bytes
eload	Load binary dump to emulator memory
esave	Save emulator memory to binary file
wipe	Wipe a LEGIC Prime tag

Chameleon

The chameleon is not technically a cloner, although the hardware could clone, the firmware does not allow for this. This is more of an emulation or simulation tool, but the good news is that you don't need to waste RFID tags and it works great with UID and doesn't require a special magic card.

Command	Description
CHARGING?	Returns if the battery is currently being charged (TRUE) or not (FALSE)
HELP	Returns a comma-separated list of all commands supported by the current firmware
RESET	Reboots the Chameleon, i.e., power down and subsequent power-up. Note: A reset usually requires a new Terminal session.
RSSI?	Returns the voltage measured at the antenna of the Chameleon, e.g., to detect the presence of an RF field or compare the field strength of different RFID readers.
SYSTICK?	Returns the system tick value in ms. Note: An overflow occurs every 65,536 ms.
UPGRADE	Sets the Chameleon into firmware upgrade mode (DFU). This command can be used instead of holding the

	RBUTTON while power-on to trigger the bootloader.
VERSION?	Requests version information of the current firmware
Button Commands	
RBUTTON=?	Returns a comma-separated list of supported actions for pressing the right button shortly.
RBUTTON?	Returns the currently set action for pressing the right button shortly. DEFAULT: SETTING_CHANGE
RBUTTON=<NAME>	Sets the action for pressing the right button shortly.
LBUTTON=?	Returns a comma-separated list of supported actions for pressing the left button shortly.
LBUTTON?	Returns the currently set action for pressing the left button shortly. DEFAULT: RECALL_MEM
LBUTTON=<NAME>	Sets the action for pressing the left button shortly.
RBUTTON_LONG=?	Returns a comma-separated list of supported actions for pressing the right button a long time.
RBUTTON_LONG?	Returns the currently set action for pressing the right button a long time. DEFAULT: SETTING_CHANGE
RBUTTON_LONG=<NAME>	Sets the action for pressing the right button a long time.
LBUTTON_LONG=?	Returns a comma-separated list of supported actions for

	pressing the left button a long time.
LBUTTON_LONG?	Returns the currently set action for pressing the left button a long time. DEFAULT: RECALL_MEM
LBUTTON_LONG=<NAME>	Sets the action for pressing the left button a long time.
LED Commands	
LEDGREEN=?	Returns a comma-separated list of supported events for illuminating the green LED
LEDGREEN?	Returns the currently set event for lighting the green LED
LEDGREEN=<NAME>	Sets the event for which the green LED is lit. DEFAULT: POWERED
LEDRED=?	Returns a comma-separated list of supported events for illuminating the red LED
LEDRED?	Returns the currently set event for lighting the red LED
LEDRED=<NAME>	Sets the event for which the green LED is lit. DEFAULT: SETTING_CHANGE
Log Commands	
LOGMODE=?	Returns a comma-separated list of supported log modes
LOGMODE?	Returns the current state of the log mode
LOGMODE=<NAME>	Sets the current log mode. DEFAULT = OFF
LOGMEM?	Returns the remaining free space for logging data to the SRAM (max. 2048 byte)

LOGDOWNLOAD	Waits for an XModem connection and then downloads the binary log - including any log data in FRAM.
LOGCLEAR	Clears the log memory (SRAM and FRAM)
LOGSTORE	Writes the current log from SRAM to FRAM and clears the SRAM log.

The following commands influence the currently selected slot only:

Command	Description
CONFIG=?	Returns a comma-separated list of all supported configurations
CONFIG?	Returns the configuration of the current slot
CONFIG=<NAME>	Sets the configuration of the surrent slot to <NAME> (See Configurations)
UIDSIZE?	Returns the UID size of the currently selected card type in Byte
UID?	Returns the UID of a card in the current slot
UID=<UID>	Sets a new UID, passed in hexadecimal notation.
READONLY?	Returns the current state of the read-only mode

READONLY=[0;1]	Activates (1) or deactivates (0) the read-only mode (Any writing to the memory is silently ignored)
MEMSIZE?	Returns the memory size occupied by the current configuration in Byte
UPLOAD	Waits for an XModem connection in order to upload a new virtualized card into the currently selected slot, with a size up to the current memory size
DOWNLOAD	Waits for an XModem connection in order to download a virtualized card with the current memory size
CLEAR	Clears the content of the current slot
STORE	Stores the content of the current slot from FRAM into the Flash memory
RECALL	Recalls/restores the content of the current slot from the Flash memory into the FRAM
TIMEOUT=?	Returns the possible number range for timeouts. See also Timeout commands.
TIMEOUT=<NUMBER>	Sets the timeout for the current slot in multiples of 128 ms. If set to zero, there is no timeout. See also Timeout commands.
TIMEOUT?	Returns the timeout for the current slot. See also Timeout commands.

Reader Commands	Using these commands only makes sense, if the slot is configured as reader. See also ISO14443A Reader Functionality
SEND <BYTEVALUE>	Adds parity bits, sends the given byte string <BYTEVALUE>, and returns the cards answer
SEND_RAW <BYTEVALUE>	Does NOT add parity bits, sends the given byte string <BYTEVALUE> and returns the cards answer
GETUID	Obtains the UID of a card that is in the range of the antenna and returns it. This command is a Timeout command.
DUMP_MFU	Reads the whole content of a Mifare Ultralight card that is in the range of the antenna and returns it. This command is a Timeout command.
IDENTIFY	Identifies the type of a card in the range of the antenna and returns it. This command is a Timeout command.
THRESHOLD=?	Returns the possible number range for the reader threshold.
THRESHOLD=<NUMBER>	Globally sets the reader threshold. The <NUMBER> influences the reader function and range. Setting a wrong value may result in malfunctioning of the reader. DEFAULT: 400

THRESHOLD?	Returns the current reader threshold.
AUTOCALIBRATE	Automatically finds a good threshold for communicating with the card that currently is on top of the Chameleon. This command is a Timeout command.
FIELD?	Returns whether (1) or not (0) the reader field is active.
FIELD=[0;1]	Enables/disables the reader field.

Commercial Off the Shelf

There are many commercially available options to clone RFID cards, most of these are going to be focused on unencrypted technologies. The most popular are listed in the next few pages of this chapter.

I-COPY X

iCopy-X is a handheld RFID card cloning
machine with the latest, simplest, and full
decode functionality. With upgrades in RFID
card technology, iCopy-X offers the most
comprehensive support and coverage among all
RFID cloning devices out in the market.

This is one of the most comprehensive
solutions available, but that does come with
a large price tag

125KHz Handheld RFID Duplicator

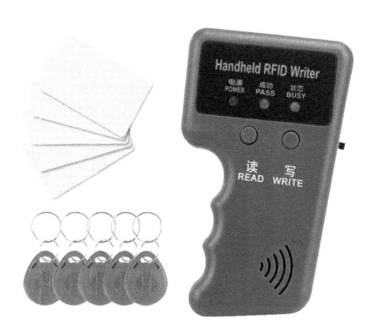

How to Use:
1. Switch on, power led flashes one second.
2. Put RFID Card in read area.
3. Press read button, read led flashes with beep sounds.
4. Put RFID card to be wrote in read area.
5. Press write button, write led flashes with beep sounds.
6. Copy done.

Card Copier Specifications:
Frequency: 125KHz
Power Supply: 2*AAA battery.it can be use 20,000 times.(Battery not included)
Reading Distance: RFID Key tag/2.5cm; RFID Card/6cm
Dimensions: Approx. 4.72*2.76*1.18in / 12*7*3cm
Package Information:
Package Size: Approx. 5.31*3.27*1.57in / 13.5*8.3*4cm
Package Weight: Approx. 3.17oz / 90g

Card Specifications:
T5577 chip which compatible with EM4305 chip
Size:3.5 x 2.5 x 0.5cm
Material: Plastic
Operation frequency: 125Khz
Read and rewrite key tags
Detecting distance: 0-5cm
Application: It applies to access control, hotel locks, staff attendance and school campus access and payment control, identification and security systems, parking lot entry and payment, social security management, transportation payment, municipal and ancillary service payment, etc.

RFID Copier Duplicator

ID/IC full hand and voice broadcast (In English)
Support read write frequency: 125KHz, 250KHz, 375KHz, 500KHz, 625khz, 750kHz, 875KHz, 1000KHZ, 13.56MHz (ISO1443A/B), support HID Prox cards.
Support card read: EM4100/EM4200, Mi-fare classic, UID cards, Ultralight, HID 1386/1326/1346, Ntag203
Support cards write: T5577, EM4305, HID 1386/1326/1346, UID cards.

High definition 3.2 inch full color screen,
clear display, easy to read data.
Full frequency band, with voice broadcast
function.
Compatible with all blank cards (T5577,
EM4305, etc.).
Full digital button input, you can directly
enter the card number.
Handheld and compact, easy to carry.
Comes decoding, can directly prepare all
kinds of intelligent cards access door.
color: White
Material: ABS
size:152 x 80 x 21mm

iCopy3

All sector decryption: super decrypt function, break through the encryption line, easy to copy.
Software upgrade: regularly upgrade software platform
Big data technology: using big data Shared

library of built-in password technology,
decryption successfully deposited in the
password database automatically, next to the
same card copy does not need to be repeated.
Automatic scanning: work frequency
from(0.100 MHZ) 100 KHZ to 13.56 MHZ
Enhanced detection: detection can be
performed to the special card, automatic
filtering false password.

Support types:
13.56MHZ IC(M1-S50), S70,TK/EM4100,H-ID16,
H-ID35/37, GID64/50/40/32/16,SID64/50/40/32
UID,FUID,CUID,ZXUID,EUID,ICUID,E5500,E5550,T
5557,EM4305,EL8265/5200/6200/7200/8200/6608/
6609

Softward description

Note:
If 13.56MHZ Mifare card encrypted, must use software unencrypted

Softeare download
https://yunpan.cn/c6C3L78QbjrkT
password: d2d2

USB 125 khz ID Card RFID
Writer/Copier/Reader/Duplicator

Support fully compatible format uem4100 ID
card (64 bit, Manchester code)
Working frequency 125 khz
Communication rate 106 Kbit / s
Serial transmission rate 9600 Kbit / s
DC power supply 5 V (±5%)
Current consumption <

100 mA effective working distance 100mm

RS232 / R232 interface

Extended I/O port stateless

Indication 2 LED (power LED status indicator), BUZ

Operating temperature - 10 ° C to 70 ° C

Storage temperature -20 ° C to 80 ° C

Maximum working humidity 0~95% relative humidity

Product size 110mm x 80mm x 26mm

Weight: 130g

Smartphone

Most modern smartphones have NFC capabilities built in and with NFC apps you can copy/clone some cards. The amount of RFID cards that are cloneable will be less than what can be accomplished with a dedicated cloner.

Brand	Model	Operating System	NFC Tags Compatible		
			MIFARE Classic® / 1k (M1K S50)	MIFARE Ultralight®	NTAG®
Apple	iPhone 6S / 6S Plus **	iOS	✗	✗	✗
Apple	iPhone SE **	iOS	✗	✗	✗
Apple	iPhone 7 / 7 Plus **	**iOS 13**	Format	✓	✓
Apple	iPhone 8 / 8 Plus **	**iOS 13**	Format	✓	✓
Apple	iPhone X **	**iOS 13**	Format	✓	✓
Apple	iPhone Xs / Xs Max **	**iOS 13**	Format	✓	✓
Apple	iPhone Xr **	**iOS 13**	Format	✓	✓
Apple	iPhone 11, 11 Pro, 11 Pro Max	**iOS 13**	Format	✓	✓
Apple	iPhone 12, 12 Pro, 12 Pro Max, 12 Mini	**iOS 14**	Format	✓	✓

Samsung	Galaxy A3 2017 ***	**Android 8.0**	✓	✓	✓
Samsung	Galaxy A5 2017 ***	**Android 8.0**	✓	✓	✓
Samsung	Galaxy A6 / A6 Plus	Android	✗	✓	✓
Samsung	Galaxy A7 / A7 2018	Android	✗	✓	✓
Samsung	Galaxy A8 / A8 2018 / A8 Plus 2018	Android	✗	✓	✓
Samsung	Galaxy A8s	Android	✗	✓	✓
Samsung	Galaxy A20e	Android	✗	✓	✓
Samsung	Galaxy A31	Android	✗	✓	✓
Samsung	*memory* Galaxy A40	Android	✓	✓	✓
Samsung	Galaxy A41	Android	✗	✓	✓
Samsung	Galaxy A42	Android	✗	✓	✓
Samsung	Galaxy A50 / 50s	Android	✗	✓	✓
Samsung	Galaxy A51 / A51 5G	Android	✗	✓	✓
Samsung	Galaxy A52 / A52 5G	Android	✗	✓	✓
Samsung	Galaxy A60	Android	✗	✓	✓
Samsung	Galaxy A70 / A70s	Android	✗	✓	✓
Samsung	Galaxy A71 / A71 5G	Android	✗	✓	✓
Samsung	Galaxy A72	Android	✗	✓	✓
Samsung	Galaxy A80	Android	✗	✓	✓
Samsung	Galaxy A90	Android	✗	✓	✓
Samsung	Galaxy Ace 3	Android	✗	✓	✓

Samsung	Galaxy Ace 4	Android	✗	✓	✓
Samsung	Galaxy Alpha	Android	✗	✓	✓
Samsung	Galaxy A Quantum	Android	✗	✓	✓
Samsung	Galaxy Core 2	Android	✗	✓	✓
Samsung	Galaxy Core Advance / Plus / Prime	Android	✓	✓	✓
Samsung	Galaxy Express 2	Android	✗	✓	✓
Samsung	Galaxy F62	Android	✗	✓	✓
Samsung	Galaxy Fame / Lite	Android	✗	✓	✓
Samsung	Galaxy Grand Duos / Prime	Android	✗	✓	✓
Samsung	Galaxy J3 (2016)	Android	✗	✓	✓
Samsung	Galaxy J4 Plus	Android	✗	✓	✓
Samsung	*memory* Galaxy J5 (2016) / (2017)	Android	✓	✓	✓
Samsung	Galaxy J5 Pro	Android	✗	✓	✓
Samsung	Galaxy J6 Plus	Android	✗	✓	✓
Samsung	Galaxy J7 (2017)	Android	✗	✓	✓
Samsung	Galaxy K Zoom	Android	✗	✓	✓
Samsung	Galaxy M21	Android	✗	✓	✓
Samsung	Galaxy M30s	Android	✗	✓	✓
Samsung	Galaxy M31	Android	✗	✓	✓
Samsung	Galaxy M51	Android	✗	✓	✓

Samsung	Galaxy M	Android	✗	✓	✓
Samsung	Galaxy Mega	Android	✗	✓	✓
Samsung	Galaxy Mini 2	Android	✓	✓	✓
Samsung	Galaxy Nexus	Android	✗	✓	✓
Samsung	Galaxy Note 2	Android	✓	✓	✓
Samsung	Galaxy Note 3 / Neo	Android	✗	✓	✓
Samsung	Galaxy Note 4	Android	✗	✓	✓
Samsung	Galaxy Note 5	Android	✗	✓	✓
Samsung	Galaxy Note 7 / 7 FE	Android	✗	✓	✓
Samsung	Galaxy Note 8	Android	✓	✓	✓
Samsung	Galaxy Note 9	Android	✓	✓	✓
Samsung	*memory* Galaxy Note 10 / 10+	Android	✓	✓	✓
Samsung	Galaxy Note 20 / 20 Ultra	Android	✗	✓	✓
Samsung	Galaxy Note Edge	Android	✗	✓	✓
Samsung	Galaxy S2 *	Android	✓	✓	✓
Samsung	Galaxy S3 / S3 Neo	Android	✓	✓	✓
Samsung	Galaxy S3 Mini *	Android	✓	✓	✓
Samsung	Galaxy S4 / S4 Active / S4 Mini / S4 Zoom	Android	✗	✓	✓
Samsung	*memory* Galaxy S5	Android	✓	✓	✓
Samsung	Galaxy S5 Mini	Android	✗	✓	✓

Samsung	Galaxy S6 / S6 Edge / S6 Edge Plus	Android	✗	✓	✓
Samsung	*memory* Galaxy S7 ***	**Android 8**	✓	✓	✓
Samsung	Galaxy S7 Edge	Android	✗	✓	✓
Samsung	*memory* Galaxy S8 ***	**Android 8**	✓	✓	✓
Samsung	Galaxy S8+	Android	✗	✓	✓
Samsung	*memory* Galaxy S9	Android	✓	✓	✓
Samsung	Galaxy S9 Plus	Android	✗	✓	✓
Samsung	Galaxy S10 / S10 Plus / S10e	Android	✗	✓	✓
Samsung	Galaxy S20	Android	✗	✓	✓
Samsung	Galaxy S21 (all version)	Android	✗	✓	✓
Samsung	Galaxy Superior	Android	✓	✓	✓
Samsung	*tablet_androidmemory* Galaxy Tab Active / Pro	Android	✓	✓	✓
Samsung	Galaxy Xcover 3 (G388F) / (G389F)	Android	✓	✓	✓
Samsung	Galaxy Xcover 4	Android	✓	✓	✓
Samsung	Galaxy Xcover 4s	Android	✗	✓	✓
Samsung	Galaxy XCover FieldPro	Android	✗	✓	✓
Samsung	Galaxy Young	Android	✓	✓	✓

Samsung	Galaxy Z Flip	Android	✗	✓	✓
Samsung	Galaxy Z Fold 2	Android	✗	✓	✓
Samsung	W2019	Android	✗	✓	✓
Samsung	W21	Android	✗	✓	✓
Google	Galaxy Nexus	Android	✓	✓	✓
Google	Nexus 10	Android	✗	✓	✓
Google	Nexus 4	Android	✗	✓	✓
Google	Nexus 5	Android	✗	✓	✓
Google	*memory* Nexus 5X	Android	✓	✓	✓
Google	Nexus 6	Android	✗	✓	✓
Google	*memory* Nexus 6P	Android	✓	✓	✓
Google	Nexus 7 (2012) / (2013)	Android	✗	✓	✓
Google	Nexus 9	Android	✗	✓	✓
Google	Nexus S	Android	✓	✓	✓
Google	Pixel / XL	Android	✓	✓	✓
Google	Pixel 2 / XL	Android	✓	✓	✓
Google	Pixel 3 / XL / 3a / 3a XL	Android	✗	✓	✓
Google	Pixel 4 / XL	Android	✗	✓	✓

The word **"Format"** means that the NFC Tag must be formatted in NDEF format, to be read by this model.

Block NFC

RFID blocking cards, shields, and protectors work in several ways. They are either passive or active. Passive shields or protectors can absorb the RFID signal or deflect it. Active RFID shields or protectors use a microchip. They typically send out an interfering signal. It is effectively pretending to be another card causing a card clash issue in the reader. Or it drains the power from the transmitting signal required to power the chip in your card. If the chip can't receive enough power to function it can't transmit a reply.

Passive shielding works in two ways, reflective loss (bounced radio frequency energy) and absorption loss (energy/power absorbed in the shield). Both types reduce the transmitting rf signal that would power the RFID chip.

However, deflecting or absorbing RFID signals for banking cards and contactless payment cards at 13.56 MHz is far simpler than blocking security cards and passes at 125 kHz. 125 kHz requires near field shielding to protect the RFID chip.

Sleeves

RFID blocking sleeves are great way to block RFID chips for individual cards, but are cumbersome if you have multiple cards.

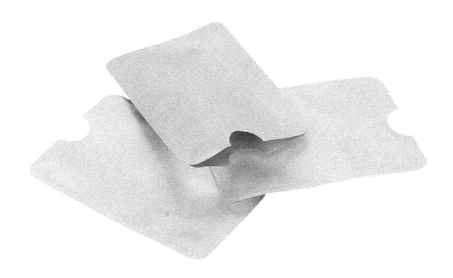

Wallet

RFID blocking wallets are great if you have multiple cards that you are needing to protect. Wallets are typically accessories that we have with us everywhere and with a RFID blocking wallet you are protected from attacks while walking around.

Disable RFID

The technology behind RFID is simple, it is simply a transistor with an antenna attached. If you create a large enough electromagnetic field, the RFID chip will overload and melt, rendering it useless.

DIY

//Warning//

The below instructions are for educational purposes only. There is risk of shock and could be hazardous to your health. You assume all risk by following the below guide below. This is not intended to be used for any other purpose than to educate about RFID overload techniques.

//Warning //

Supplies: Disposable camera, speaker wire or
CAT5 cable. Extra: Soldering Iron, Switch

1. Remove the plastic cover and battery
   cover

2. Separate the case to reveal the circuit board

3. Remove the circuit board from camera
   casing

4. Identify circuit board parts and
   functions

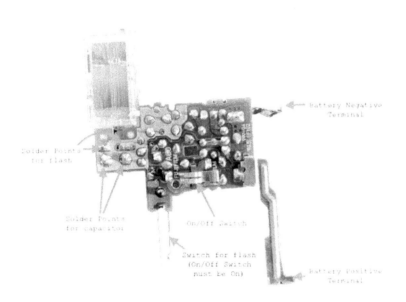

5. Make a coil. I used an old CAT5 cable,
   I seperated the orange pair and made a
   coil using hot glue to get it rigid.

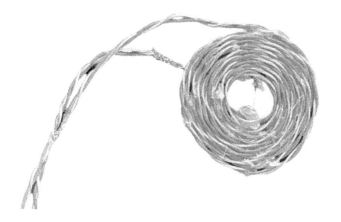

6. Charge the capacitor by putting battery in place, and turning the On/Off switch to On
7. Put RFID tag on the coil and touch each lead from the coil to one of the capacitors terminals.

8. Optional: Remove the flash and put the coil leads in place of the flash

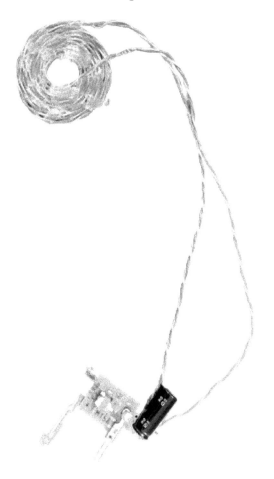

NFCKILL

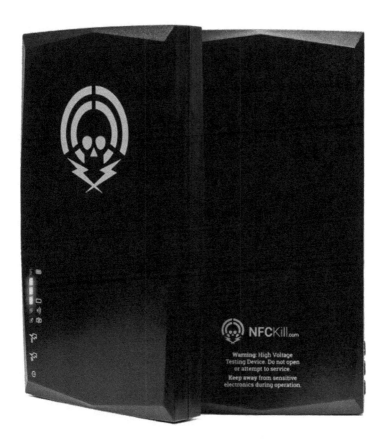

While the NFCKill is tuned to cover the most
common Low and High Frequencies of RFID:
125KHz - 13.56MHz. Likewise, it is able to
inductively couple with most devices that
contain a form of coil.

Targeted Frequencies:

- High Frequency (13.56MHz)
  - Tags: All known tags (MIFARE Family, HID iCLASS, Calypso, Contactless Payment, etc.)
  - Hardware: Most reader and writer hardware, NFC-equipped phones, Contactless Payment Terminals, etc.
  - Effective Range: 0 - 6cm
  - Throughput: 50 cards per cycle (max 6000 per minute)
- Low Frequency (125 - 134KHz)
  - Tags: All known tags (HID, Indala, etc.)
  - Hardware: Most reader and writer hardware
  - Effective Range: 0 - 5cm
  - Throughput: 1 card per cycle (max 120 per minute)
- Ultra-High Frequency (850 - 930 MHz)
  - UHF technology is fundamentally different and requires a different device. The **NFCKill UHF** is purpose built for UHF tags.

NFCKILL UHF

NFCKill UHF Specifications

- Dimensions: 245 x 85 x 80 mm
- Antenna Size: 160 x 150mm
- Weight: 2.5KG
- Voltage: 10 - 14VDC
- Current: 6A (Max)
- Instantaneous Power: 15kW
- Operating Range: 5cm
- Pulse Frequency: 5Hz
- Device Operating Lifecycle: > 1,000,000 cycles

Compatible Tags:

- All passive UHF RFID tags
- UHF RFID Tags: 800 - 960MHz
- Gen1 / Gen2
- 98% coverage of common antenna formats

Notes

NOTES:

NOTES:

NOTES:

NOTES:

NOTES:

NOTES:

NOTES:

NOTES:

NOTES:

NOTES:

NOTES:

NOTES: